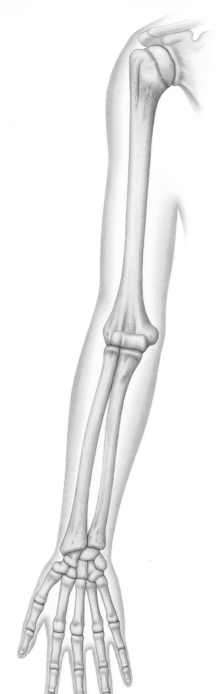

The Human Body

The Skeleton and Muscular System

Carol Ballard

RSVP
RAINTREE
STECK-VAUGHN
PUBLISHERS
The Steck-Vaughn Company

Austin, Texas

TITLES IN THE SERIES

The Heart and Circulatory System
The Stomach and Digestive System
The Brain and Nervous System
The Lungs and Respiratory System
The Skeleton and Muscular System
The Reproductive System

Published by Raintree Steck-Vaughn Publishers,
an imprint of Steck-Vaughn Company

Library of Congress Cataloging-in-Publication Data
Ballard, Carol.
The skeleton and muscular system / Carol Ballard.
p.   cm.—(The human body)
Includes bibliographical references and index.
Summary: Explains the various parts of the human skeleton and
different types of muscles and their functions.
ISBN 0-8172-4805-6
1. Musculoskeletal system—Juvenile literature.
[1. Muscular system. 2. Skeleton.]
I. Title. II. Series: Human body (Austin, Tex.)
QP301.B316   1997
612.7—dc21   96-29688

Printed in Italy. Bound in the United States.
1 2 3 4 5 6 7 8 9 0 02 01 00 99 98

Consultant: Dr. Tony Smith, Associate Editor of the British Medical Journal

Picture Acknowledgments
The publishers would like to thank the following for use of their photographs:
Eye Ubiquitous 43; Science Photo Library 6, 12, 14, 15,
16, 23, 32, 33, 34, 42; Wayland Picture Library 37, 44, 45.

CONTENTS

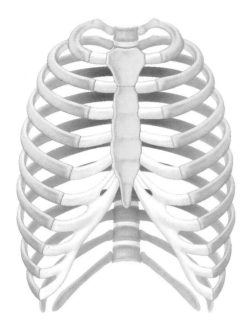

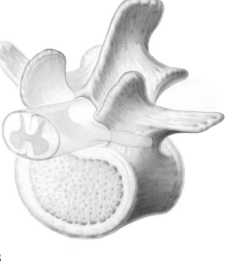

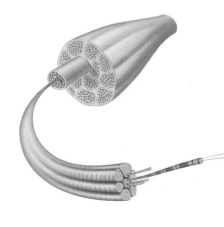

Introduction

Without bones and muscles, our bodies would be soft and unable to move, just like enormous jellyfish. The bones of the skeleton provide support and protection for the body. The **joints** where bones meet allow flexibility and movement. The muscular system provides a mechanism for moving the bones.

Some bones of the skeleton form a strong framework that supports the body and protects important organs. Find out more on page 10.

Other bones are specially designed for movement. The bones in the arms and hands allow us to lift and carry and to perform tiny, precise movements. The bones in the legs and feet help us to run and jump. Find out more on page 22.

The junction where two bones meet is called a joint. Bones are held together at joints by tough fibers called **ligaments.** Different types of joints allow different types of movement. Find out more on page 28.

Muscles can pull bones, allowing an enormous range of movement. Some simple movements involve only a pair of muscles, while more complex movements involve many muscles. Find out more on page 40.

Bones and muscles are very important, so it makes sense to take care of them by getting exercise, by eating the right food, and by learning how to avoid injury. Find out more on page 44.

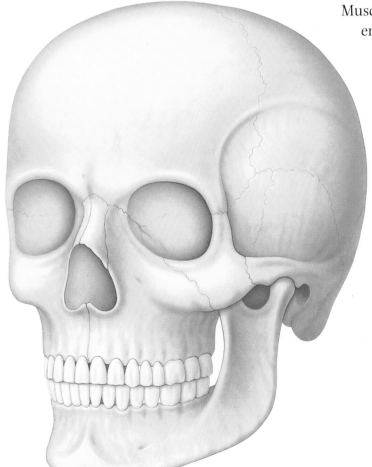

◀ **The bones of the skull protect the brain.**

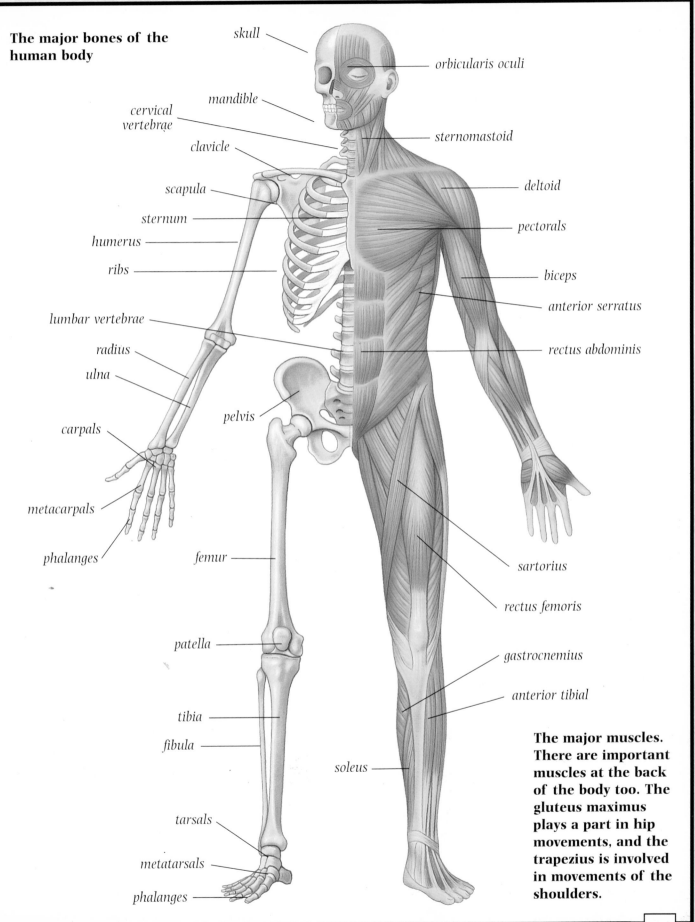

The major bones of the human body

skull

orbicularis oculi

mandible

cervical vertebrae

sternomastoid

clavicle

scapula

deltoid

sternum

pectorals

humerus

ribs

biceps

anterior serratus

lumbar vertebrae

radius

rectus abdominis

ulna

pelvis

carpals

metacarpals

phalanges

femur

sartorius

rectus femoris

patella

gastrocnemius

anterior tibial

tibia

fibula

soleus

The major muscles. There are important muscles at the back of the body too. The gluteus maximus plays a part in hip movements, and the trapezius is involved in movements of the shoulders.

tarsals

metatarsals

phalanges

Bones

Although bones are hard and strong, they are made of living **tissue**. Every bone has nerves and a blood supply, and bones grow and develop with the rest of the body.

The bones of the human skeleton all have the same basic structure, but their shapes are adapted to suit their different functions. Although each bone has its own special name, bones can be grouped together according to their shape and size.

The bones of the arms and legs are longer than their width and are called long bones. They are good at acting as levers and are important in moving the body.

The length, width, and thickness of short bones are all about the same. These small, odd-shaped bones are found in the wrists and ankles.

Flat bones are thin, curved bones. The ribs, scapula (shoulder blade), and sternum (breastbone) are all flat bones. The large, flat surfaces of these bones allow strong muscles to be attached to them. The flat bones of the skull form a protective shell for the brain.

Irregular bones are exactly what their name suggests and do not have a regular shape. The bones of the face and of the **spine** are all irregular bones.

Some joints have small bones inside that help to make the joint work efficiently. These are called **sesamoid bones** because they can look like sesame seeds. One example is the patella (kneecap).

As a baby develops, its bones fuse. Sometimes the bones do not fuse completely, so some people have tiny, extra bones. These are called accessory bones, and they usually occur in the feet. On X rays, they sometimes look like broken bones.

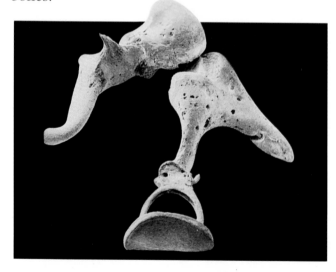

▲   The tiny hammer, anvil, and stirrup bones inside the ear are the smallest bones in the body. They have been magnified several times in this picture.

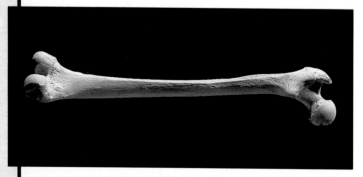

◄ The longest bone in the body is the femur.

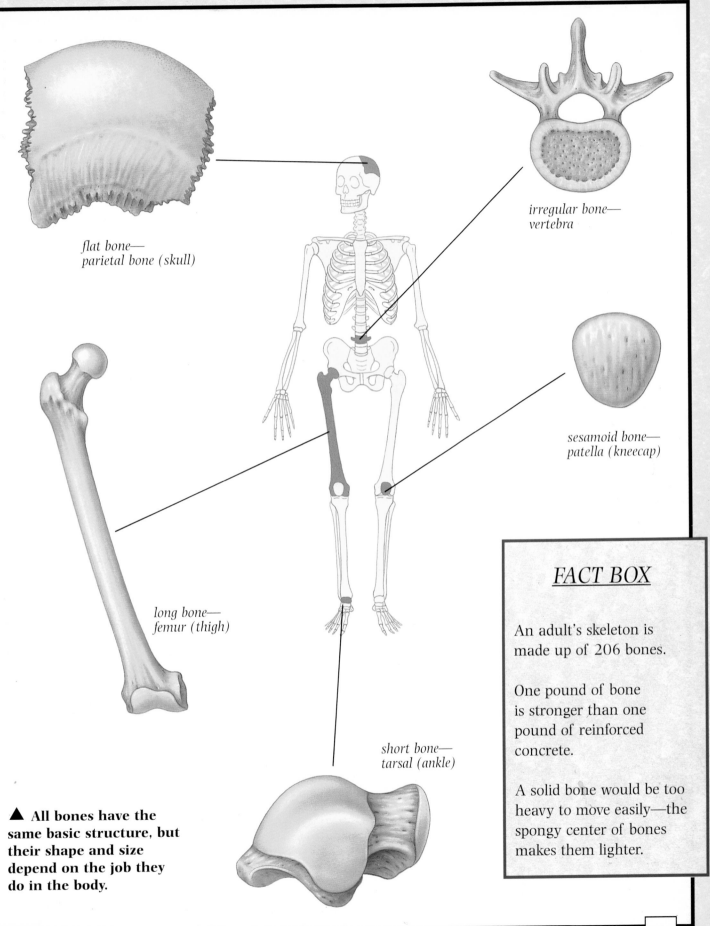

flat bone—
parietal bone (skull)

irregular bone—
vertebra

sesamoid bone—
patella (kneecap)

long bone—
femur (thigh)

short bone—
tarsal (ankle)

▲ **All bones have the
same basic structure, but
their shape and size
depend on the job they
do in the body.**

FACT BOX

An adult's skeleton is
made up of 206 bones.

One pound of bone
is stronger than one
pound of reinforced
concrete.

A solid bone would be too
heavy to move easily—the
spongy center of bones
makes them lighter.

Inside a Bone

Bones are not as solid as they look. They are made up of several layers, designed to make them as strong and light as possible. Some long bones, such as those in the arms and legs, are hollow in the middle, for lightness. Bones are living tissue, so they need a good blood supply to bring **nutrients** and oxygen to them and to take away waste products. They also need **nerves** to send information to the brain about pain or damage caused to a bone.

About 50 percent of a bone is water. The rest is **minerals** and protein. Minerals, mainly calcium and phosphorus, help to strengthen the bones. A tough, stringy protein called collagen runs through the bones.

Bone tissue is constantly being remodeled and renewed. There are three main types of cells in bones, each with a special function:
1. Osteoblasts make new bone for growth and to repair any damage.

2. Osteocytes keep bones healthy by carrying nutrients and waste products to and from blood vessels.
3. Osteoclasts destroy bone tissue, in order to release minerals into the blood for use in other parts of the body.

The functions of these three types of cells are balanced so that bones stay healthy and strong.

Most bones have an outer covering, or "skin," called the periosteum. This is a tough **membrane** made of fibers. It contains nerves and blood vessels. It also contains special cells (osteoblasts) that help to repair the bone after a fracture.

Inside the periosteum is a layer of hard material called **compact bone**. This looks somewhat like the ivory of an elephant's tusk and is very strong. It has holes and channels running through it that carry blood vessels and nerves to the inner layers of the bone.

The center of the bone is made of **spongy bone**. This is a network of tiny pieces of bone called trabeculae, joined together to form a mesh, somewhat like a honeycomb. The spaces between the trabeculae are filled with a soft jelly called **bone marrow**. There are two types of bone marrow. Red marrow produces red blood cells and is found at the ends of bones.

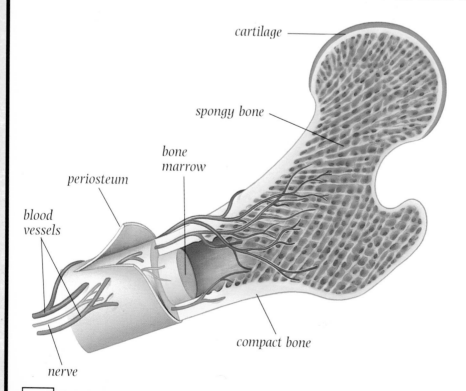

cartilage

spongy bone

bone marrow

periosteum

blood vessels

nerve

compact bone

◀ **A bone is made up of layers of different material, as this cross section shows.**

▼ How bones develop

1. Cartilage forms.

2. Minerals are deposited, making the cartilage stronger and harder.

3. Bone tissue begins to develop at the center of the cartilage.

4. Blood vessels supply nutrients to the developing bone.

5. The bone grows bigger.

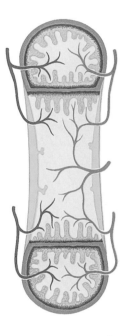

6. The bone now grows from the growth plates at the ends of the bone.

Yellow marrow is mainly fat and is found in the central hollow spaces of long bones.

Bones develop from a rubbery material called **cartilage**, which forms in a baby before birth. As the baby develops, minerals are deposited in the cartilage, making it harder and stronger. Bone tissue begins to develop at the center of the cartilage, and blood vessels carry nutrients to the developing bone. As more bone tissue is formed, the bone grows longer. Eventually, the center of the bone is fully formed.

A child's bones are soft, but they gradually become harder and stronger as more minerals are deposited. This hardening process is called ossification. As a child grows, new bone tissue is made at the ends of the bones, from special areas called growth plates.

Some cartilage remains at the ends of the bones to protect them. In other places, cartilage remains throughout life and does not turn into bone—for example, the nose is stiffened by cartilage, not bone.

What Do Bones Do?

The bones of the skeleton have three main physical functions: support, protection, and movement. Bones are also involved in the production of some blood cells, and they play an important role in the storage and release of some minerals.

Bones support every part of the body. They form a strong framework that enables us to stand upright.

The bones of the skull are fused, forming a strong, bony case that protects the brain. The ribs form a cage that is attached to the sternum (breastbone) at the front of the body and to the spine at the back. The heart and lungs lie inside this cage, which gives them good protection. Each bone of the spine (vertebra) has a hole through its center, and these holes are lined up to form a channel. The **spinal cord** runs through this channel, protected by the column of vertebrae. The bones of the **pelvic girdle** provide a protective cradle for the intestines and the female reproductive organs.

Individual bones are not very flexible, although they do bend very slightly, especially in children. It is the junctions between the bones that allow movement, not the bones themselves. These junctions are called joints. Bones are held together at joints by tough fibers called ligaments. Bones cannot move themselves, but they act as levers when they are pulled by muscles.

The marrow inside bones produces all the body's red blood cells. Bone marrow contains special cells called mother cells, which produce about 200 billion red blood cells every day. Bone marrow is also important in the production of some white blood cells. Like red blood cells, these are made by the special mother cells. When they leave the bone marrow, white blood cells are not all fully developed and may travel to other parts of the body to mature.

Bones contain large quantities of the minerals calcium and phosphorus and smaller amounts of other minerals such as magnesium and zinc. These, especially calcium, make bone hard and strong. When the body has plenty of calcium, the bones store it. If a person's diet does not contain enough calcium, the bones will release some into the blood and the blood carries it to other parts of the body that need it. If the calcium released from the bones is not replaced, the bones gradually become weaker and more brittle.

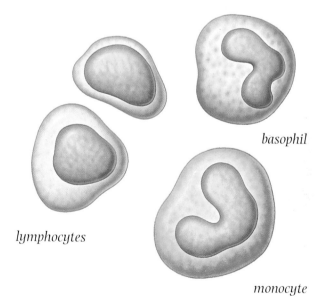

basophil

lymphocytes

monocyte

White blood cells

▲ ▶ **Several types of white blood cells are produced by the red bone marrow.**

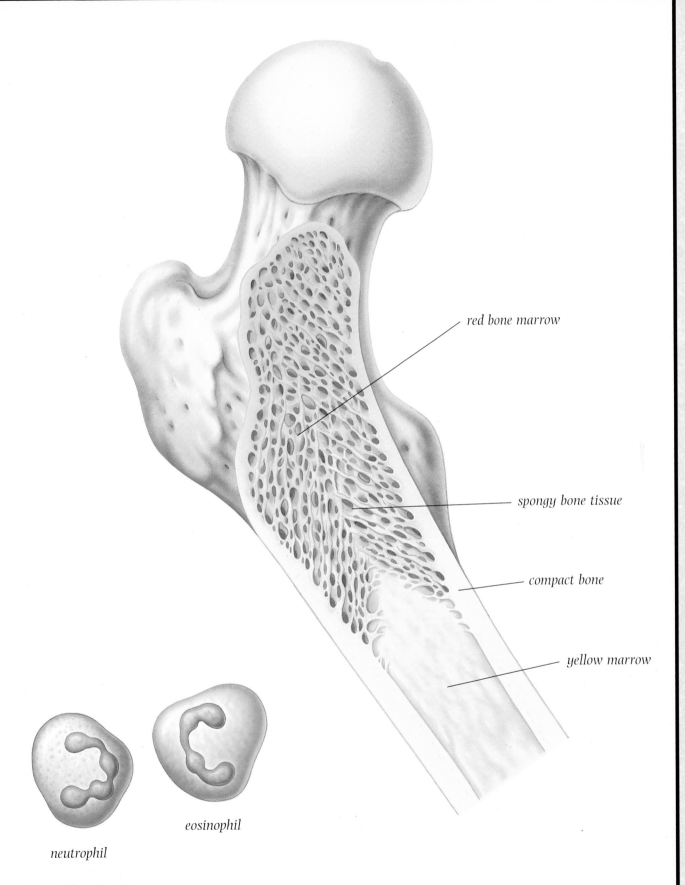

red bone marrow

spongy bone tissue

compact bone

yellow marrow

neutrophil

eosinophil

White blood cells

Bone Diseases and Problems

Bones are strong but they can break, usually as a result of an accident. Children's bones are supple, so they may bend rather than break, and they heal quite quickly. As people get older, however, their bones become weaker and more brittle, so their bones break more easily and take much longer to mend.

A broken bone is called a **fracture**. There are different types of fractures. The simplest fractures mend quickly, but more serious, complicated fractures can take a long time to heal. One or more operations might be needed to ensure that the break mends properly.

If a bone is thought to be fractured, it must be prevented from moving. A leg can be kept still with a rigid support called a splint, and the patient can then be carried on a stretcher. An arm can be kept still and supported in a sling.

Doctors usually take an X ray so that they can see the exact position and type of fracture. This helps them decide what treatment is needed. Some bones, such as ribs, heal quickly without any treatment, although the patient usually has to avoid strenuous exercise for a while. A simple fracture of an arm or a leg may be treated by wrapping the limb in wet plaster bandage. This hardens as it dries, making a strong plaster cast. It keeps the pieces of bone still and in the right position until the fracture has healed. An operation might be needed to position the bones properly if the fracture is a complicated one. Metal plates can be screwed into the bone to make it stronger, and a plaster cast then keeps the bone still until the fracture has healed.

When a bone breaks, blood vessels inside the bone are torn. Blood escapes and forms a clot at the site of the fracture. New blood vessels and bone tissue slowly begin to grow. This holds the ends of the bone together. A ring of cartilage develops around the outside of the fracture. The cartilage is gradually replaced by new spongy bone. Eventually, new compact bone forms and the bone is fully healed. A small bump is usually the only remaining sign that the bone has been fractured.

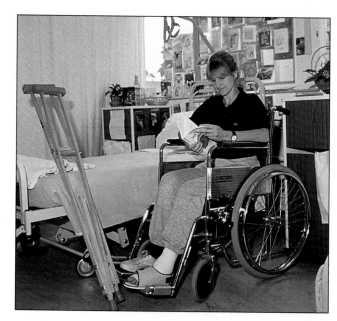

Different types of fractures

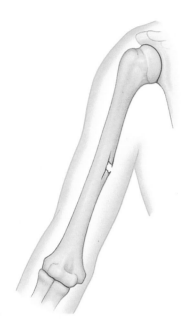

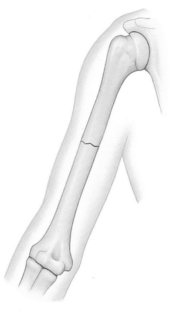

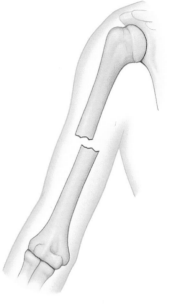

"Greenstick" fractures are common in children. One side of the bone is broken but the other side is only bent.

In a simple fracture, the bone is broken into two pieces, but they stay in line and remain inside the skin.

In a complete fracture the bone is broken into two pieces that move out of line but do not pierce the skin.

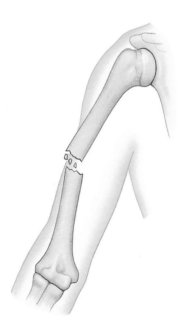

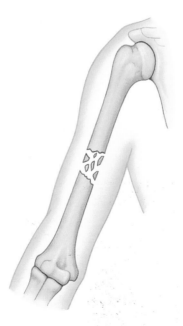

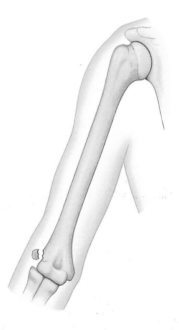

In a compound fracture, the bone is broken into two pieces that pierce the skin.

In a comminuted fracture, the bone is shattered into many small pieces.

Sometimes, a small piece of bone is broken off the main bone. This is a chip.

◄ **Once the fractured bone has been encased in plaster, the patient can get around more easily. A wheelchair helps the patient move without putting weight on the broken bone.**

Bone Diseases

A fracture is the most common bone problem. There are some diseases that make bones more likely to fracture, and other conditions that can stop bones from developing properly or making them grow abnormally.

Some children suffer from an inherited disease called osteogenesis imperfecta. Their bones do not contain enough protein and minerals to make them strong, so they break very easily. Sometimes, a baby may even be born with broken bones. As the child grows up, the bones gradually become stronger and less likely to break.

As people get older, they may suffer from osteoporosis. This means that their bones become weak and fragile, and even ordinary movements can cause a bone to break. Elderly people who have apparently fallen and broken a hip have often done just the opposite—the hip has broken as they walked, causing them to fall. The bones of people with osteoporosis also crumble a little and squash together. Many elderly people are shorter than they were when they were younger, because their vertebrae get packed together more tightly.

Vitamins, especially vitamin D, are important for maintaining healthy bones, and a lack of vitamin D can cause a disease called rickets. The bones of children with rickets do not contain enough minerals, so they are weak and do not grow properly.

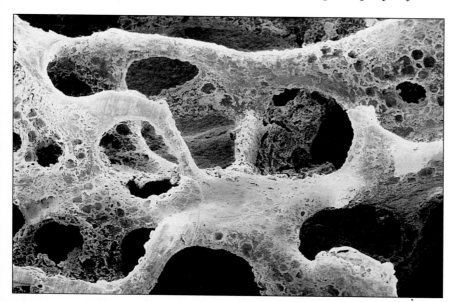

◀ In osteoporosis, the spongy bone tissue becomes less dense. On this vertebra, some of the spongy bone has thinned to nothing and the empty areas show up as small, round, dark patches. This picture was taken through a microscope.

This X-ray image shows the bone deformities in the legs of a child suffering from rickets. The bones are bowed and weak. Healthier diets and more plentiful food mean that rickets is now quite rare in developed countries.

The leg bones can be too weak to bear the weight of the child when he or she is standing, so they gradually bend, causing bowlegs and knock-knees. Rickets is much rarer today than it used to be, because people now tend to eat much healthier food. Cleaner, smog-free air and better living conditions have helped too, because sunlight is an important source of vitamin D.

The bone marrow at the center of a bone can sometimes be infected by bacteria, which can be carried in the blood or may enter the bone after an injury. Bone infections can usually be cured quickly with antibiotics.

Bone marrow can be used to treat diseases because it plays an important role in the development of white blood cells. Some people have diseases of their immune systems, which means that they do not have enough white blood cells working properly. One way of treating these diseases is by a bone marrow transplant. Bone marrow from a healthy person is removed and transplanted into the patient, and the white blood cells in the healthy bone marrow often help the patient's own immune system to recover.

The Head

The bony structure that forms the head is called the skull, and it is made up of two main parts: the cranium and the face.

The cranium is the rounded, hollow part of the head, which protects the brain. It is made up of eight bones joined together by zigzag joints called sutures. The main bones of the cranium are the parietal bone (at the top and side), the occipital bone (at the back), and the frontal bone (the forehead).

When a baby is born, there are soft spaces, called fontanels, between the bones of the cranium. Gradually, during the first year of life, these gaps close and the bones fuse. Because the sutures have zigzag edges, the bones lock tightly together and cannot move.

FACT BOX

There are twenty-nine bones in the head—fourteen in the face, eight in the cranium, six inside the ears, and one at the top of the throat.

Men's skull bones are usually thicker than women's.

Bony supports inside the skull help it to withstand the enormous pressures exerted when the teeth chew.

The fourteen bones of the face make up the front of the skull. Thirteen of the bones are fused and cannot move. The lower jaw (the mandible) is hinged at each side. This allows it to move up and down so that the mouth can open and close. It also has a limited amount of side-to-side movement. Teeth are embedded in the upper jaw (the maxilla) and in the lower jaw.

Each eye lies protected in a strong, bony hollow called the orbital cavity. The cheeks are shaped by the zygomatic bones. The bridge of the nose is made of two small, oblong bones at the top of the nasal cavity. The rest of the nose is supported and stiffened by cartilage, not bone.

Some of the bones of the face have hollow spaces inside them, which are filled with air. These spaces are called sinuses. They allow sounds from the vocal cords to "echo" inside the skull, making the voice louder. Having sinuses also makes the skull lighter than it would be if all the bones were solid.

Inside each ear are three tiny bones—the ossicles—called the malleus (hammer), incus (anvil), and stapes (stirrup). They get their names from their strange shapes. The ossicles form a chain of levers, carrying sound waves from the eardrum to the inner ear.

The hyoid bone is not attached to any other bones, but is held in place by muscles and ligaments. It supports the tongue, so it is important in speaking and swallowing.

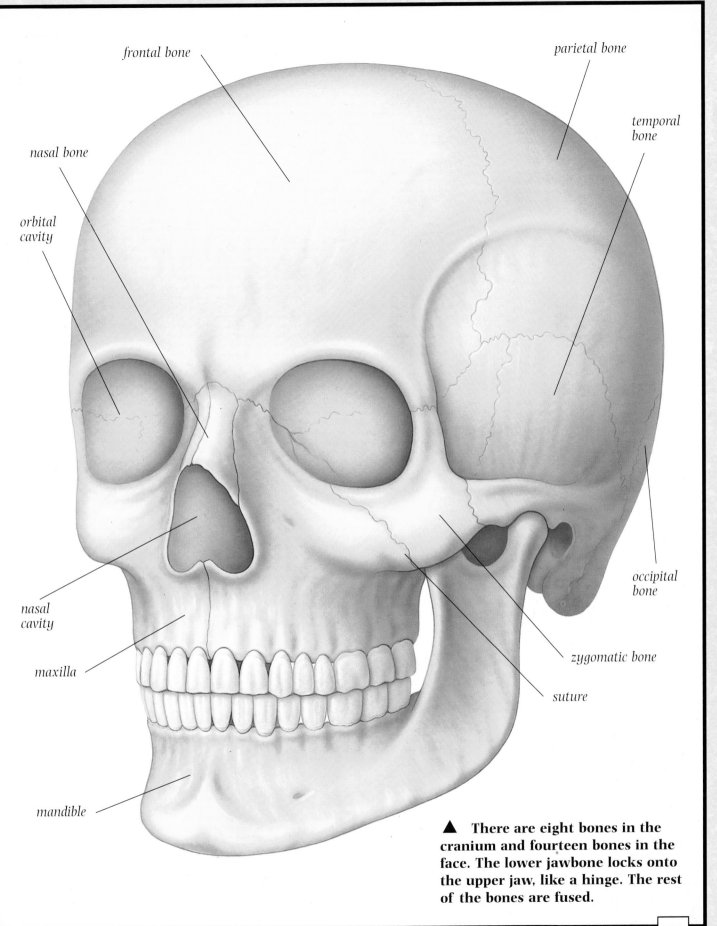

frontal bone

parietal bone

temporal bone

nasal bone

orbital cavity

occipital bone

nasal cavity

zygomatic bone

maxilla

suture

mandible

▲ There are eight bones in the cranium and fourteen bones in the face. The lower jawbone locks onto the upper jaw, like a hinge. The rest of the bones are fused.

The Spine

The spine acts as a central support for the rest of the body, allowing us to stand upright. Its slight "S" shape ensures that the weight of the body is evenly balanced. Bones, muscles, cartilage, ligaments, and **tendons** all work together to make the spine strong and flexible. The spine allows movement from side to side, backward and forward, and up and down. It also allows twisting movements.

The spine contains thirty-three individual bones, called vertebrae. These are divided into five groups: the cervical (neck), thoracic (upper back), lumbar (lower back), sacral (pelvic), and "tail" vertebrae. The five sacral vertebrae are fused to form one bone called the sacrum. The four tail vertebrae are also fused, making one bone called the coccyx.

Disks of cartilage lie between each vertebra. These disks act as shock absorbers, cushioning and protecting the bones and allowing them to move without grinding against each other. The disks of cartilage make up about a quarter of the length of the spine. Most people are slightly shorter at the end of the day than they are in the morning —during the day, the weight of the body gradually squashes and flattens the disks. The disks return to their correct shape and size overnight.

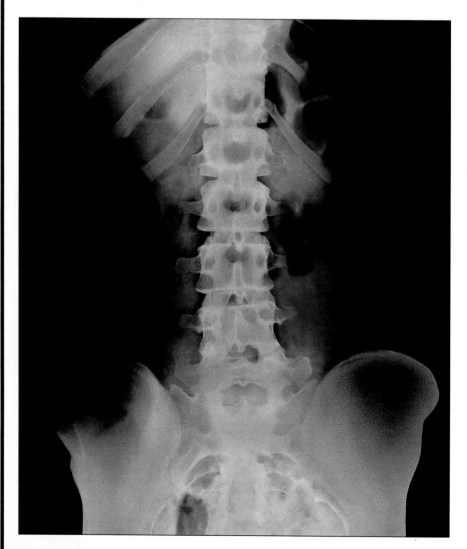

◄ **In this X-ray image of the spine, the lower thoracic vertebrae (with ribs attached) can be seen at the top. The five lumbar vertebrae are in the center, with the top of the sacrum just below them.**

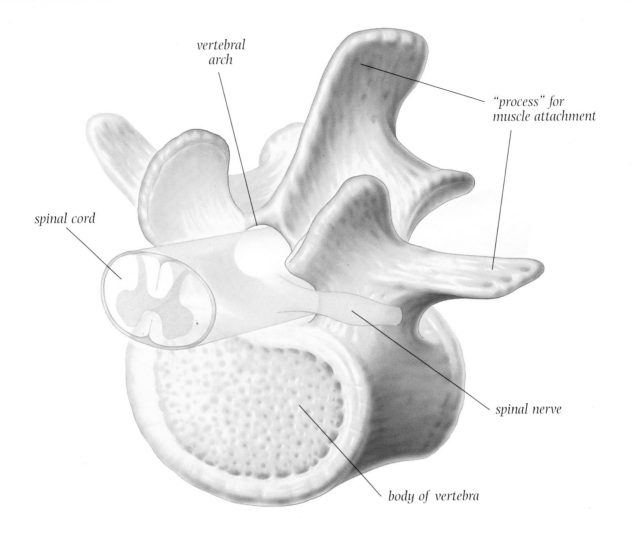

vertebral
arch

"process" for
muscle attachment

spinal cord

spinal nerve

body of vertebra

Vertebrae are irregularly shaped bones, but their general structure is similar. The "body" of a vertebra is a roughly cylindrical shape, with flat top and bottom surfaces. It supports the weight of the body. A bony spike, called a "process," juts out at the back of each vertebra and another process juts out at each side. Muscles are attached to them and they act as levers, helping the muscles to move the vertebrae. In addition to these processes, there are two others that jut out at each side and these link with the processes of the vertebrae above and below. This keeps the vertebrae in the correct position.

An arch of bone (the vertebral arch), together with the body of the vertebra, forms a strong,

▲ **Each vertebra encloses the spinal cord in a strong, bony ring.**

bony ring that surrounds and protects an important bundle of nerves called the spinal cord.

The top two vertebrae are different from the rest, and their special design allows the head to move. The first vertebra is the atlas, which supports the head. It is a wide vertebra, but it has no process at the back and no body. The joint between the atlas and the second (axis) vertebra allows the head to nod up and down. This joint also acts as a pivot, allowing the head to move from side to side.

The Chest

The chest (thorax) is made up of the bones of the upper spine (the thoracic vertebrae), the breastbone (sternum), and the ribs, together with muscles and cartilage. These form a strong structure called the rib cage. This protects the heart and lungs and provides support for the bones of the shoulders. It also plays an essential part in breathing.

Most people have twelve pairs of ribs, although some people have eleven or thirteen pairs. Each rib is a flat bone, curved into a flattened semicircle. The ribs are joined to the vertebrae at the back. The top seven pairs (ribs one to seven) are joined to the sternum at the front by strips of cartilage. The next three pairs (eight to ten) are only indirectly attached to the sternum: a strip of cartilage connects these ribs together and joins them to rib seven. The bottom two pairs (eleven and twelve) are called floating ribs because they are not attached to the sternum at all.

The sternum is a flat, straight bone, shaped like a dagger. It is about 6 in. (15 cm) long in an adult. It runs down the center of the chest and provides an anchorage for the ribs. The clavicles (collarbones) also are joined to the sternum.

There are three sections to the sternum. The upper section has a notch on each side, where the clavicles attach. The joint between the upper and middle section acts as a hinge, allowing the sternum to move during breathing. The bottom section is the smallest.

In children it is made of cartilage, but it gradually changes into bone. Several important muscles and ligaments are attached to this section.

The spaces between the ribs are called intercostal spaces. The intercostal muscles that lie in these spaces connect each rib to the ribs above and below. These muscles are very important in breathing. When the intercostal muscles contract, they pull each rib upward and outward. This creates more space inside the rib cage, so air is drawn into the lungs. When the intercostal muscles relax, the ribs move downward and inward. This reduces the space inside the rib cage, and air is forced out of the lungs.

A sheet of muscle called the diaphragm also helps to control the space inside the rib cage. The diaphragm is attached to the cartilage of the rib cage. When it is relaxed, the diaphragm arches upward into the rib-cage space. When it contracts, it pulls downward, increasing the space inside the rib cage.

▶ **The rib cage is a strong structure that gives good protection to the heart and lungs. It is also flexible, so it can expand and contract as we breathe.**

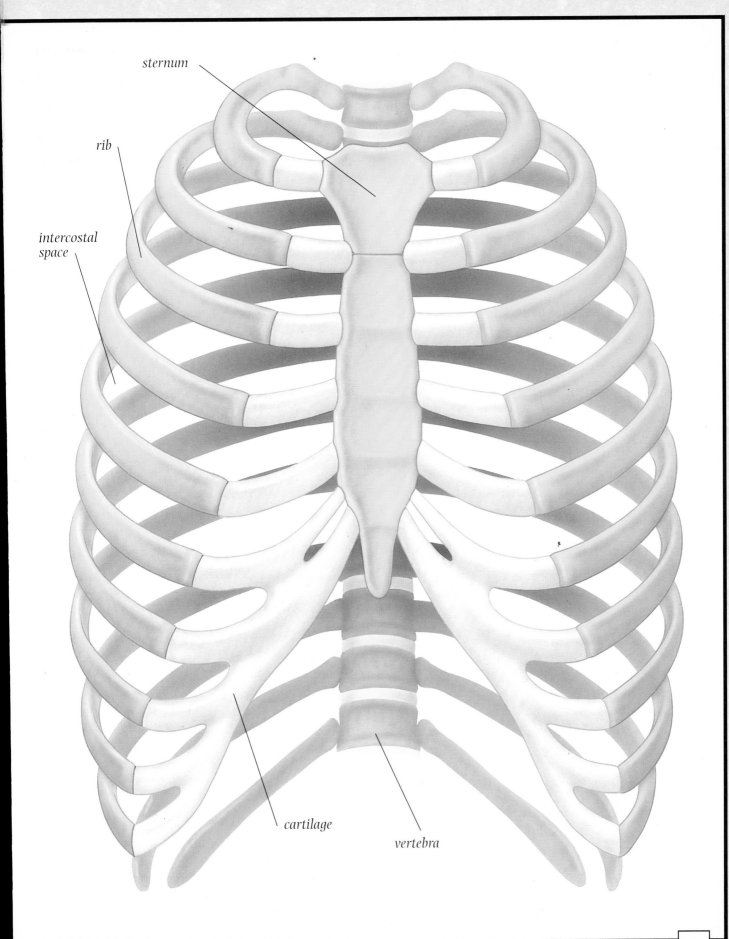

sternum

rib

intercostal
space

cartilage

vertebra

The Limbs

The arms and hands have the same basic design as the legs and feet.

The scapulae (shoulder blades) and clavicles (collarbones) form a horizontal barlike structure called the **pectoral girdle**. Each clavicle is joined to the top of the sternum (breastbone) and is held in place by strong ligaments, which allow limited movement. There are flexible joints between the clavicles and the scapulae. Powerful muscles that move the arms are attached to the scapulae.

The arms are attached to the pectoral girdle at the shoulder joints. The clavicles hold the shoulder joint away from the rib cage so that the arms can swing freely. The clavicles are quite thin and are in a very vulnerable position; a broken collarbone is one of the most common bone injuries, and often happens as a result of an accident while playing sports.

There are three bones in each arm. The humerus is the long upper bone between the shoulder and the elbow. The head of the humerus is rounded and fits into a hollow in the scapula. At the lower end of the humerus is the elbow, the joint between the humerus and the lower arm. The lower arm contains two long bones, the **radius** and the **ulna.** The ulna is longer than the radius and is on the outside of the arm, the same side as the little finger. The radius is on the side closest to the body, the thumb side. These two bones can rotate, allowing the wrist to twist. If you hit your elbow on something, you might say you have hit your "funny bone." In fact, you have hit a nerve that runs over the end of the ulna.

At the lower end of the radius and ulna are the eight short bones (carpals) of the wrist. These are arranged in two rows of four bones. The carpals are connected to each other by ligaments, which also hold them in place and restrict their movements. Each carpal can move only slightly in relation to the other carpals, but together these tiny movements allow a lot of flexibility.

The palm of the hand is made up of five long bones called metacarpals. These are arranged like a fan, spreading out from the wrist. The fourteen finger bones (phalanges) are connected to the hand bones at the knuckles. Each finger contains three phalanges, but the thumb contains only two.

The human hand is capable of very precise, accurate movements. The thumb is long and very mobile, and is able to touch each of the fingertips. It allows us to hold objects very precisely in a "pincer grip" that is unique to humans and other primates.

The bones of the shoulder, arm, and hand together make a versatile limb. It can move in different ways and lift heavy weights. It also allows us to carry out very precise actions, such as writing. ▶

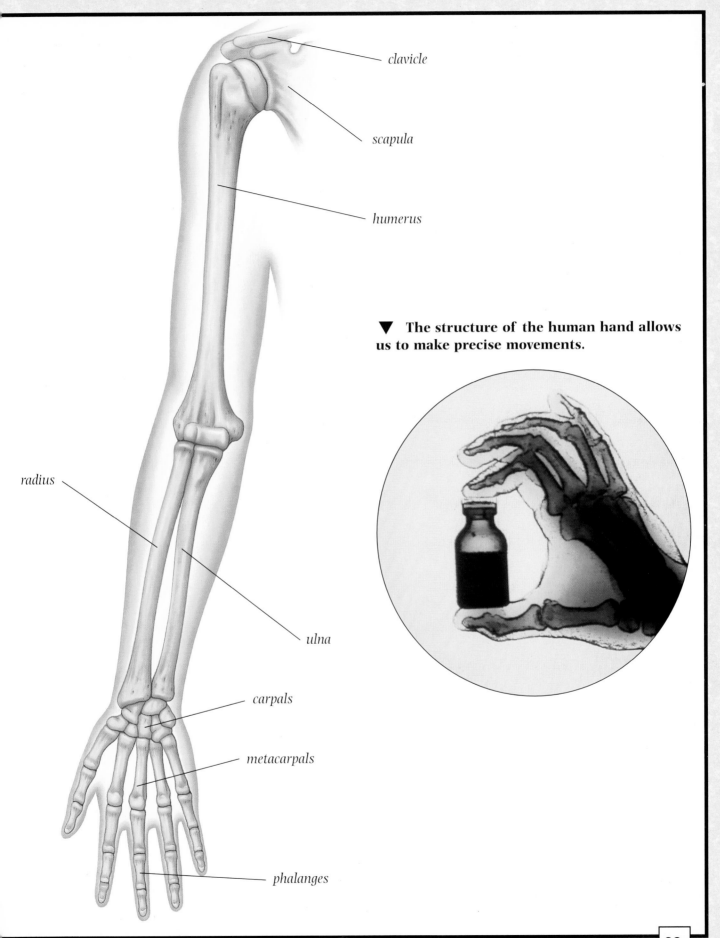

clavicle

scapula

humerus

▼ **The structure of the human hand allows us to make precise movements.**

radius

ulna

carpals

metacarpals

phalanges

Legs and Feet

The pelvis is a deep, irregular ring of bone made up of the pelvic girdle at the front and sides, and the sacrum and coccyx at the back. Strong, powerful muscles are attached to the pelvis, which is held together by ligaments. The pelvis plays an important part in transferring body weight as we move around, and it also protects the internal organs that lie within it.

The legs are attached to the rest of the body at the pelvic girdle. The pelvic girdle is made up of the right and left hip bones. In children, each hip bone is actually three separate bones: the ilium, ischium, and pubis. These gradually fuse, forming a single bone by the time a child is between fifteen and seventeen years of age.

The thigh bone is called the femur. It is the longest bone in the body and is usually also the strongest and heaviest. It is fairly smooth, apart from a ridge at the back to which muscles are attached. The femur supports the weight of the body when standing, and the joint between the femur and the pelvic girdle allows movement such as walking, running, and jumping.

At the lower end of the femur is the knee joint, which allows the lower leg to move backward and forward. The knee joint is protected by a sesamoid bone—the kneecap, or "patella." The patella lies inside a tendon. As the lower leg is bent or straightened, the patella slides along a groove at the bottom of the femur.

There are two bones in the lower leg, the tibia and the fibula. The tibia (shin bone) is much stronger and sturdier than the fibula and carries most of the body's weight. It is connected by tendons to the lower end of the femur and to the patella. The slender fibula is not strong enough to support the weight of the

---

FACT BOX

A person's height is usually about four times the length of his or her femur.

The patella is the largest sesamoid bone in the body.

The arches of the feet act as shock absorbers—they flatten out when we run.

Archaeologists can tell the sex of a skeleton from the pelvis—a woman's is usually wider and lower than a man's.

---

body. It is not connected to the knee joint, but it is important to the structure of the ankle.

At the lower end of the tibia and fibula are the seven short bones of the ankle and heel. These are called the tarsals, and they are held firmly in place by strong tendons. The foot contains five long bones, the metatarsals. These, together with the tarsals, are arranged to form an arched shape. This shape is strong and absorbs the impact of the body's weight every time a step is taken.

There are fourteen long bones, or "phalanges," in the toes of each foot—two in the big toe and three in each of the other toes. They are much shorter than the finger bones and are not usually capable of precise movements, but they play an important part in helping us to balance. Some people who cannot use their hands because of injury or birth defects learn to draw and write with their feet.

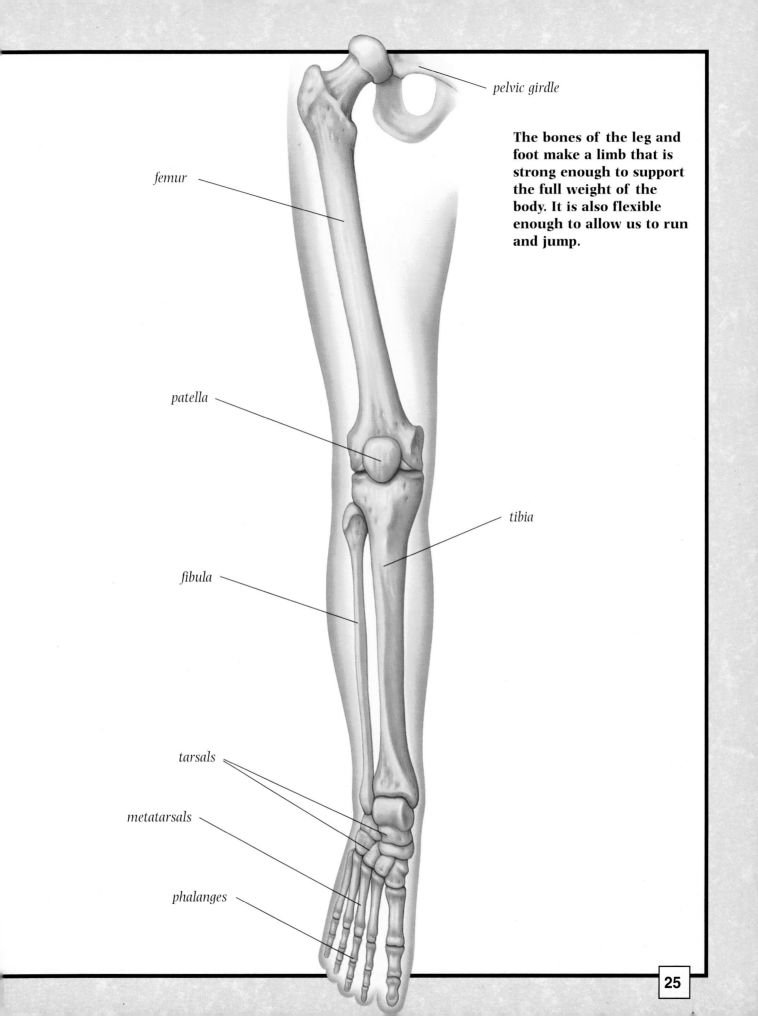

femur

patella

fibula

tarsals

metatarsals

phalanges

pelvic girdle

The bones of the leg and foot make a limb that is strong enough to support the full weight of the body. It is also flexible enough to allow us to run and jump.

tibia

Skeletal Diseases and Problems

The most common problem affecting bones is a fracture (see page 12), but other problems can also occur.

In the months before birth, a baby's bones develop and grow. Many, including the bones of the skull, gradually fuse. If the bones that form the roof of the mouth (the palate) do not fuse properly, the baby may be born with a gap there, and the inside of the mouth and nose might not be fully separated. This is called a "cleft palate." If the gap is at the front of the mouth, the lip might be separated too—a "hare-lip." Doctors can usually correct the defect by surgery.

Spina bifida is also the result of the failure of bones to develop properly before birth. If one or more vertebrae do not form a full ring, the spinal cord is not fully enclosed. The nerves of the spinal cord might protrude through the skin, protected only by a thin membrane. If this happens, the spinal cord can easily be damaged or infected.

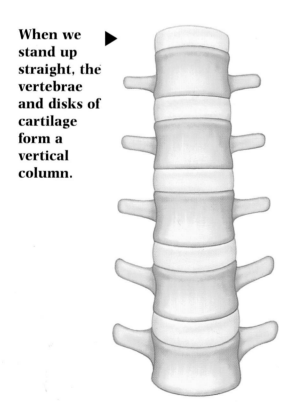

▶ **When we stand up straight, the vertebrae and disks of cartilage form a vertical column.**

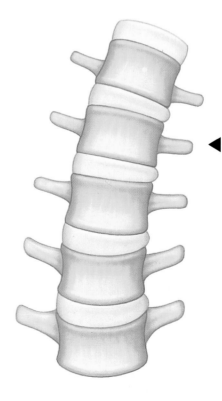

◀ **Bending to one side squashes the disks unevenly. The disks allow the spine to bend without the vertebrae rubbing against each other.**

▼ **The effect of movement on the disks of cartilage**

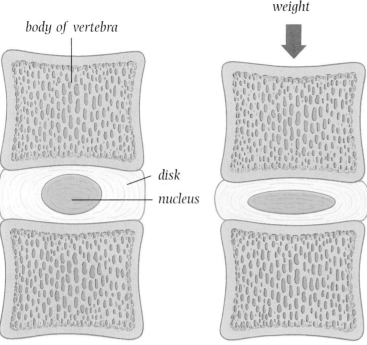
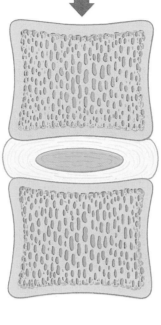
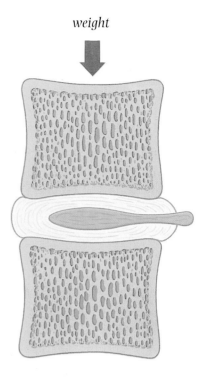
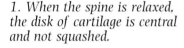

body of vertebra

weight

weight

disk

nucleus

1. When the spine is relaxed, the disk of cartilage is central and not squashed.

2. The weight of the body when standing squashes the disk of cartilage.

3. When the disks are squashed unevenly, part of a disk can jut out—this is a "slipped disk."

If the bones of the skull fuse too early during development, there might not be enough room for the brain to grow properly. This very rare condition is called microcephaly. It can be treated by an operation to widen the sutures and increase the space inside the skull.

The spine usually curves slightly in an "S" shape, but sometimes this curve is distorted. If the upper spine curves backward abnormally, a hunchback shape, called kyphosis, might develop. Lordosis is a condition in which the lower spine curves forward abnormally. The most common abnormality is scoliosis, where an area of the spine curves to the right or left side, instead of running vertically. There are many causes of abnormal curvatures of the spine, including the collapse of the vertebrae, a disease called tuberculosis of the spine,

infection during development before birth, or simply poor posture.

When we stand, the weight of the body pulls on the spine, squashing the disks of cartilage between the vertebrae. Sometimes, a disk can be squashed unevenly, so that part of it juts out beyond the line of the vertebrae. This is called a "slipped disk." It is more common in men than in women and is usually the result of bending or lifting awkwardly. It can be very painful if the disk presses on one of the spinal nerves. A range of treatments can be used, including bed rest, gentle exercise, or heat treatment. Traction is also sometimes used, which involves using a system of weights and pulleys to pull steadily on the affected area. If these treatments do not work, doctors might perform an operation to remove the damaged disk.

Joints

Joints, or "articulations," occur where two bones meet. They make the skeleton flexible, allowing individual parts to move. Without movable joints, the skeleton would be rigid and immobile.

Different parts of the body need to be able to move in different ways. The amount and direction of movement that each joint allows depends on its structure. Joints can be named and grouped according to their structure and the type of movement they permit.

Fibrous joints hold the bones tightly together, usually allowing no movement. The bones of the skull are held together by fibrous joints called sutures, and fibrous joints also hold the teeth into the jawbone. Some fibrous joints allow a small amount of movement—for example, the radius and the ulna in the lower arm are held together by a fibrous joint, but they are able to twist slightly.

In cartilaginous joints, the bones are separated by a disk of cartilage. These joints allow a little movement. The joints between the vertebrae are cartilaginous joints. Each vertebra can move just a little in relation to the vertebrae above and below it. All these little movements put together give the spine its overall flexibility.

Most of the joints in the body are **synovial joints**. These allow more movement than fibrous and cartilaginous joints and have a more complicated structure.

The end of each bone at a synovial joint is protected by a layer of cartilage. Some joints also have a disk of cartilage between the bones to act as a shock absorber. The joint is surrounded by a strong bag called the articular capsule. The inner lining of this is the synovial membrane, which produces a liquid called synovial fluid. This liquid fills the narrow space (the synovial cavity) between the two bones. It acts like oil, lubricating the joint to reduce friction and helping the bones to move smoothly.

The joint is held together by tough fibers called ligaments, which are thickenings of the articular capsule. Ligaments are strong but they are not elastic—they will tear rather than stretch if they are put under too much strain.

Inside some synovial joints are small sacs called **bursae.** These are filled with synovial fluid and they help to reduce friction inside the joint. They also allow muscles to move more easily over the surfaces of the bones.

Tendons attach muscles to bones. At some joints, long tendons are protected by tendon sheaths. These are long, cylindrical sacs filled with synovial fluid. They reduce friction and allow the tendons to slide smoothly over the bones.

Synovial joints allow more movement than ▶
other joints. This is a cross section of the
synovial joint in the knee.

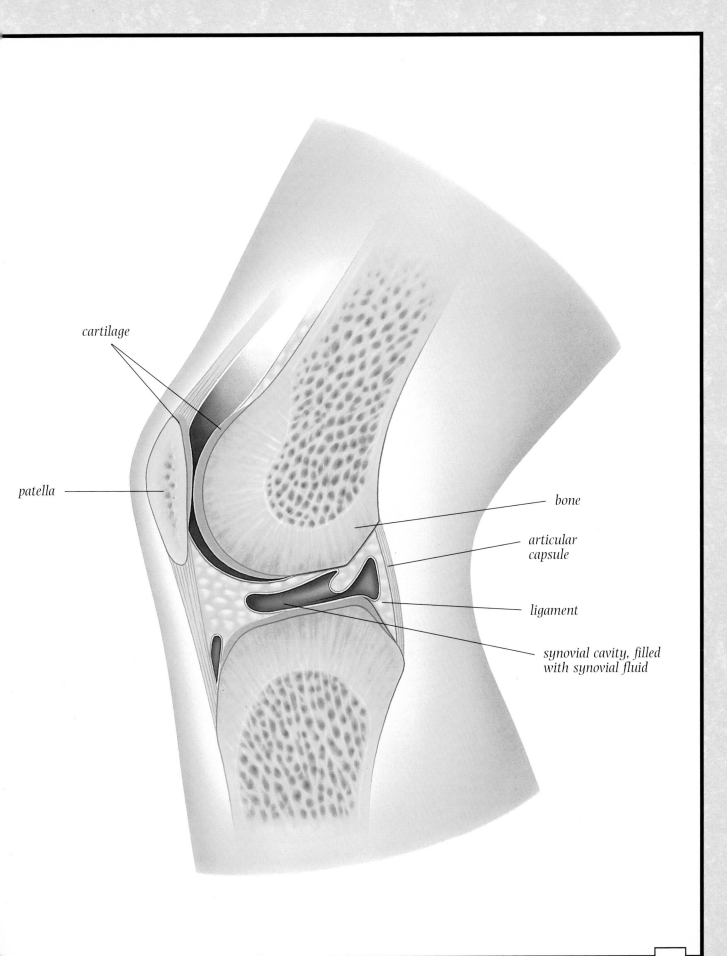

cartilage

patella

bone

articular
capsule

ligament

synovial cavity, filled
with synovial fluid

How Joints Move

Movable joints are usually synovial joints. The basic structure of each joint is the same, but the amount and direction of movement the joint allows depends on the shape of the bones involved and the way they are held together.

The simplest joints are hinge joints. These allow movement in only one direction, like the hinges that allow a door to open and close. The elbow joint is a hinge joint: the rounded end of the humerus fits into a hollow formed by the radius and ulna. The joint allows the arm to bend and straighten. Other hinge joints include the knee, the ankle, and the joints in the fingers and toes. A hinge joint also allows the head to nod up and down.

Pivot joints also allow movement in only one direction. The joint between the top two vertebrae of the spine is a pivot joint. The ring of bone of the **atlas vertebra** fits over a spike of bone on the **axis vertebra**. The joint allows the atlas vertebra to rotate, so that the head can be turned from side to side.

Condyloid joints allow movement in two directions. The wrists are condyloid joints. The ends of the radius and the ulna make a hollow into which the carpals (wrist bones) fit. The carpal bones can rock up and down and from side to side. Strong ligaments hold the carpals together so that they cannot rotate. The knuckles are also condyloid joints.

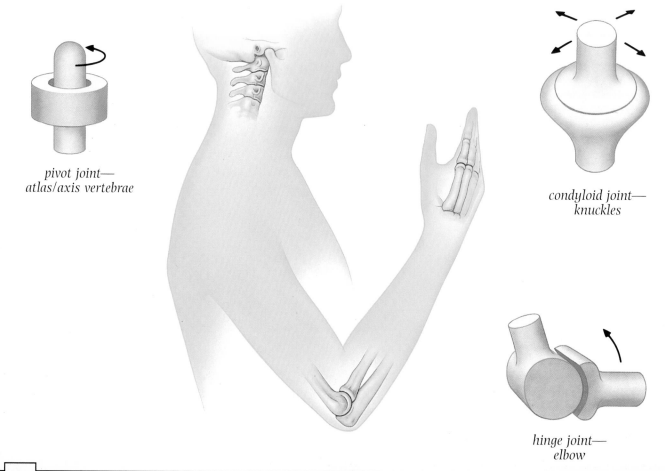

pivot joint—
atlas/axis vertebrae

condyloid joint—
knuckles

hinge joint—
elbow

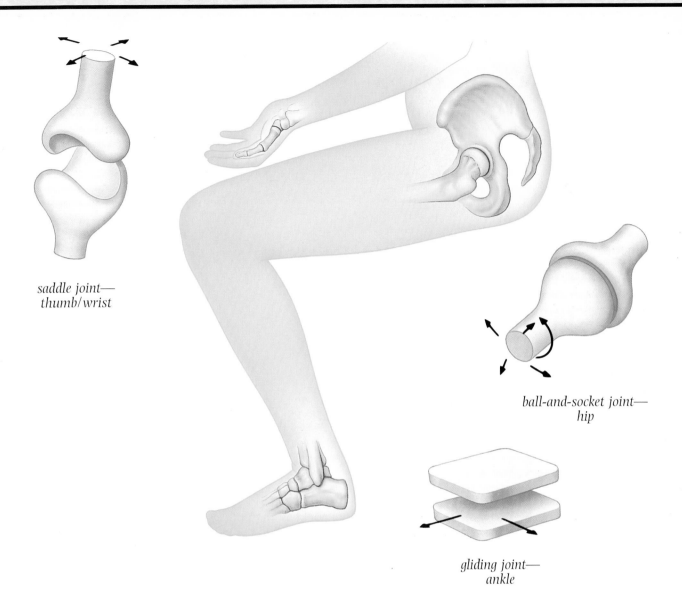

*saddle joint—
thumb/wrist*

*ball-and-socket joint—
hip*

*gliding joint—
ankle*

◀ ▲ **Different types of joints allow the body to move in different ways.**

Gliding joints are small and flat. They allow one bone to glide smoothly over the surface of another. They allow only limited movements in one direction. Examples are the joints between the carpal bones in the wrist and the tarsal bones in the ankle.

In saddle joints, the end of each bone is shaped like a saddle and the bones can move very freely. The joint between the thumb and the wrist is a saddle joint, and it allows the thumb

more freedom of movement than the knuckles allow the fingers.

Ball-and-socket joints allow more movement than other joints. The hip joint, between the femur and the pelvic girdle, is a ball-and-socket joint. The top of the femur is ball-shaped and fits into a cup-shaped socket in the pelvic girdle. The ball can rotate in the socket, allowing movement in any direction. Strong ligaments hold the bone in place, preventing extreme movements that would result in damage. The joint between the upper arm and shoulder is also a ball-and-socket joint, allowing the arm to swing freely in all directions.

Joint Diseases and Problems

Damage to joints can be the result of injury or disease. The damage might be minor and quickly mended, but in some cases the effects are serious and long term.

The most common joint injury is a sprained ankle, in which a sudden twist of the ankle tears the ligaments that hold the joint together. This can be very painful and movement of the joint is restricted until the ligaments are healed. Other joints, such as the wrist, can also be sprained. Sprains are usually treated by being kept absolutely still to rest the joint. Severe sprains sometimes need to be surgically repaired.

If the two bones at a joint become separated, the joint is said to be dislocated. The most common dislocation is at the shoulder, where the top of the humerus moves out of the socket of the shoulder. A dislocated joint can usually be put back into place by **manipulation.** Once the joint is back in place, it has to be rested and kept still for some time.

Bursitis is an inflammation of the bursae close to a joint, which makes the joint swollen and painful. It can be caused by repeated pressure placed on the bursae by actions such as bending. It is sometimes called "student's elbow" or "housemaid's knee"—names that relate the problem to the action involved. Resting the joint usually relieves the pain and allows the inflammation to go down.

Arthritis is the name for inflammation of any joint. Although people of all ages may be affected, arthritis is most common in elderly people. There are many different kinds: the most common are gouty arthritis, osteoarthritis, and rheumatoid arthritis.

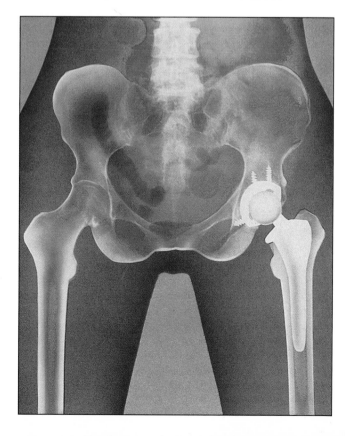

Many people have arthritic hips removed and replaced with artificial hips. The artificial hip joint can be seen on the right in this X-ray image. ▶

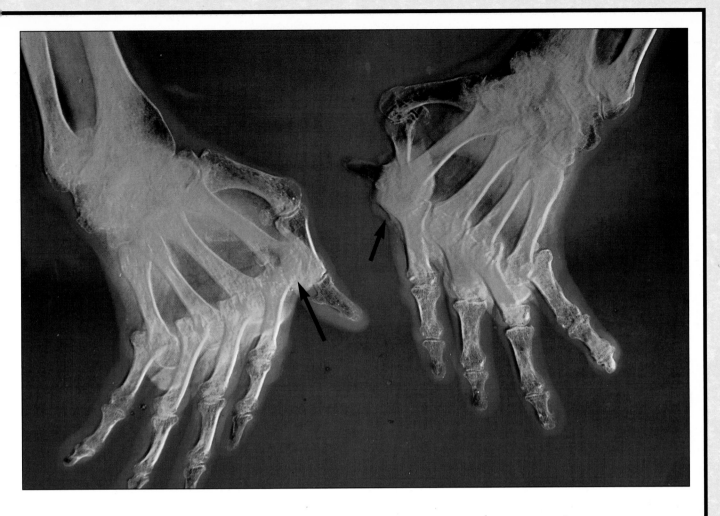

▲ **This X ray of arthritis of the hands shows the inflammation caused to the joints in the fingers by rheumatoid arthritis.**

Gouty arthritis, usually known simply as "gout," can cause inflammation and pain in any joint, but it most commonly affects the big toe. It occurs when there is too much uric acid in the body, because the excess acid forms sharp crystals in soft body tissues. Gout can usually be controlled with drugs and by eating carefully and avoiding certain acidic foods.

Osteoarthritis usually begins as part of the ageing process and is the result of joints wearing out. The surfaces of the bones become roughened and can no longer slide smoothly across each other. Eventually they rub together, making the joint stiff and painful. Joints that carry the weight of the body, such as the hip and the knee, are often affected, and this makes it difficult and painful for the person to move around. Surgeons can now remove arthritic hips and knees and replace them with artificial joints, allowing the person to move freely again.

Rheumatoid arthritis is a disease in which white blood cells attack the cartilage in the joints. It can affect any joint, but fingers, wrists, and knees are usually the first to be affected. The cartilage is eventually destroyed and the joint becomes deformed and painful. In rare cases, the bones can even fuse, preventing all movement at the joint. Rheumatoid arthritis can be treated with a variety of drugs. Although these will not cure the disease, they can reduce the pain and also slow the rate at which the joints become damaged.

Muscles

There are three main types of muscles in the body: skeletal muscle, smooth muscle, and cardiac muscle. Each type of muscle has a different structure and is designed to carry out a specific function.

Skeletal muscle is the type of muscle that is used in normal, everyday movements. It is also called voluntary muscle, because we can control it—for example, you can decide whether or not to lift your arm or nod your head.

Skeletal muscle is made up of many long, thin strands called muscle fibers. Most muscle fibers are about a little over an inch (3 cm) long, but some may be as long as 12 in. (30 cm). The fibers are bound together by a thin membrane to form bundles called fascicles. These bundles of fibers are held together by tissue that contains blood vessels and nerves. The muscle is surrounded by another layer of tissue. This outer layer forms a tough cord, called a tendon, at each end of the muscle. Most skeletal muscles are attached to bones by tendons.

▼ **This diagram shows the structure of skeletal muscle.**

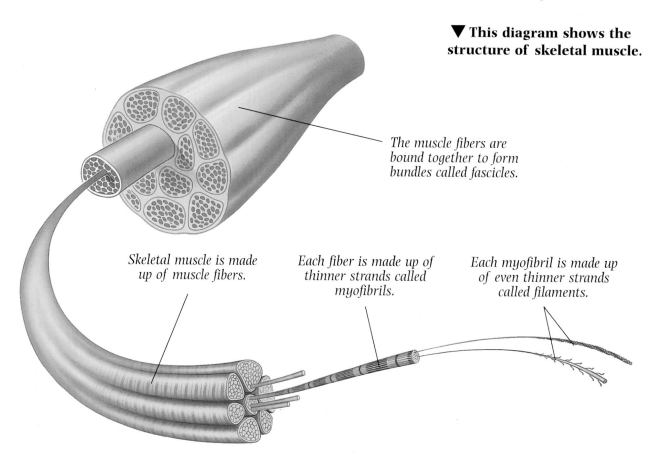

The muscle fibers are bound together to form bundles called fascicles.

Skeletal muscle is made up of muscle fibers.

Each fiber is made up of thinner strands called myofibrils.

Each myofibril is made up of even thinner strands called filaments.

Skeletal muscle is also ▶ called striated or "striped," muscle, because its fibers are made of light and dark stripes. The stripes can be clearly seen in this photograph, which was taken through a microscope.

Smooth muscle is found in internal organs and in the walls of arteries. It plays an important role in moving food through the digestive system and in controlling the blood flow through the arteries. Smooth muscle is also called involuntary muscle because most smooth muscle cannot be consciously controlled.

Smooth muscle is made up of layers of muscle tissue rather than individual muscle fibers. It can also form a ring of muscle called a sphincter. Sphincters control the size of an opening, for example, the iris in the eye and the opening from the foodpipe to the stomach.

Cardiac muscle is found only in the heart. It cannot be consciously controlled, so it is also called involuntary muscle. Cardiac muscle is extremely strong and can operate continuously. It is made of fibers, but the fibers are not arranged in bundles as they are in skeletal muscle. Instead, the fibers of cardiac muscle form a branching, crisscross network.

FACT BOX

Muscles can only pull bones. They cannot push.

There are 650 separate muscles in the human body.

Skeletal muscle makes up 35–45 percent of an average adult's body weight.

The largest muscle is the gluteus maximus, which controls hip movement.

The smallest muscle is the stapedius, in the inner ear. It is only .04 in. (1 mm) long.

Shapes and Sizes

Muscles can be grouped and named according to their shape, size, position, or action.

The shape and size of a skeletal muscle depend on its function. Some, such as the muscles in the buttocks, need to be large and powerful, to move long bones and to work against gravity. Others, such as the muscles inside the eye, are tiny, allowing very accurate and precise movements.

Strap muscles have a wide range of movement, but they are not very strong. For example, strap muscles are attached to the hyoid bone in the throat and to the sternum (breastbone). They do an important job when food is swallowed.

Fusiform muscles are spindle-shaped: thick in the middle with a strong tendon at each end.

The biceps muscle in the upper arm is a fusiform muscle. It is attached to the scapula (shoulder blade) and to the radius, and it allows the upper arm to be raised.

Pennate muscles are rather like feathers. They have a tendon at each end, but at one end the muscle is attached to the tendon by bundles of muscle fibers called fascicles. Muscles with fascicles in one direction, like the muscle that bends the thumb, are called unipennate. If a muscle has fascicles in two directions, like the muscle that bends the knee, it is called bipennate. Muscles that have fascicles arranged along several tendons are called multipennate. The fan-shaped multipennate muscles are very powerful. The deltoid muscle of the shoulder is a multipennate muscle; it allows the arm to be raised outward.

▼ **The shape of a muscle depends on its function.**

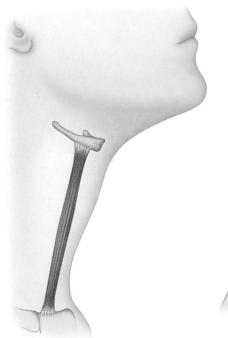

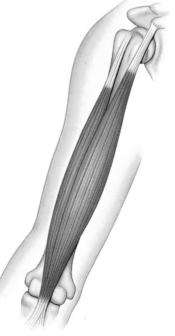

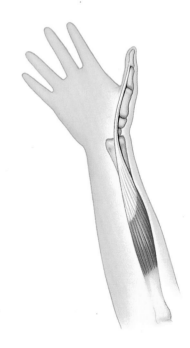

A strap muscle moves the hyoid bone to allow swallowing and speech.

A fusiform muscle raises the arm.

A unipennate muscle bends the thumb.

Circular muscles have fascicles arranged in rings around an opening. Circular muscles are important in controlling the size of an opening, such as the anus, or the iris of the eye. A ring of muscle inside the lips controls lip movement and is very important in helping us to speak. Ventriloquists try to talk without moving this muscle, but find it very difficult to produce clear speech.

Muscles in the face are attached to the skin as well as to the bones of the skull. There are more than thirty facial muscles, each with its own specialized function. A sheet of flat muscle running across the forehead lets us frown. Rings of muscle around the eyes allow us to screw up our eyes in bright light. To smile, we use muscles that run down each side of the face and can lift the corners of the mouth.

▲ **There are more than thirty facial muscles, which work together to provide us with a wide range of facial expressions.**

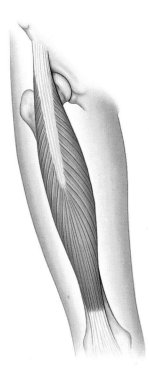

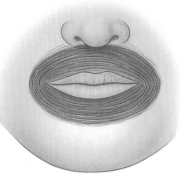

A circular muscle controls lip movements.

A bipennate muscle bends the knee.

A multipennate muscle raises the arm outward.

How Muscles Work

Muscles work by getting shorter, or "contracting." As they **contract**, they also get fatter—you can see and feel your biceps muscle getting fatter as you raise your arm.

Muscles contract in response to signals from the brain. The signals are carried by motor nerves to muscles. The nerve endings, called motor end plates, are buried in the muscle fibers. When a signal travels from the brain along the nerve, the motor end plate releases a chemical called acetylcholine. This chemical makes the muscle fibers contract.

Each muscle fiber is made up of bundles of strands called myofibrils. The myofibrils themselves are made up of even finer strands called filaments. The filaments are made of two special proteins, actin and myosin. When a muscle receives the signal to contract, the actin filaments slide past the myosin filaments. Chemical "bridges" form between the filaments, holding them in place. As they slide, the filaments are pulled closer together, so the muscle fiber gets shorter. When the muscle relaxes, the actin and myosin filaments slide past each other in the opposite direction. The spaces between the filaments get longer as the filaments move apart, so the muscle lengthens.

Muscle fibers contract in response to a signal from the brain. They stay contracted for only a very short time and will relax unless they receive another signal telling them to contract again.

To do an easy job, such as lifting a pencil, only a few fibers of the muscle need to contract. Full muscle power is needed to lift a heavy weight, so all the fibers of the muscle contract. To use the full power of a muscle, the muscle fibers are synchronized—they all receive the signal to contract at the same time.

When muscles contract, they use energy. A chemical called ATP stores energy. ATP can be changed into another chemical, ADP, and this change releases the stored energy, which can then be

relaxed myofibril

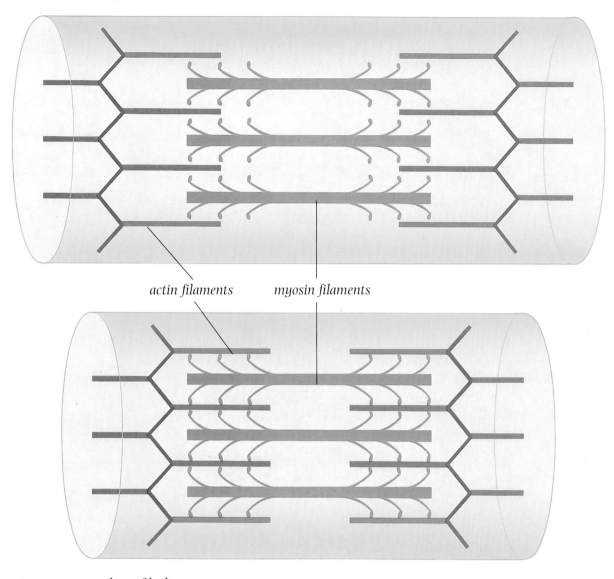

actin filaments   myosin filaments

contracted myofibril

used by the muscle. To make sure that the muscles always have a supply of energy, the body stores energy from food by changing ADP back into ATP. These chemical changes release some waste products: water, heat, carbon dioxide, and **lactic acid**. These are carried away from the muscles by the blood.

Most signals from the brain to the muscles are the result of decisions we make: you decide to pick up a book, so your brain sends signals to the appropriate muscles telling them to contract. Reflex signals are not like this. They are automatic reactions that you cannot control. Many reflex actions are protective: for example, people automatically pull their hands away from a dangerously hot object and blink if something gets into their eyes.

▲ **Actin and myosin filaments slide past each other when a muscle contracts. As they slide, the filaments are pulled closer together, making the muscle fiber shorter.**

Muscles Move Bones

Skeletal muscles and bones work together as a system of levers. The bone acts as a lever, which pivots around a joint. The muscles provide the force to move the "lever" and any weight it might be carrying.

Muscles work by contracting. They can only get shorter, so they can only pull a bone in one direction; they cannot push the bone back to its original position. Most muscles work in pairs, known as antagonistic pairs. Each muscle in an antagonistic pair has the opposite effect from the other muscle. For example, if the contraction of one muscle raises a bone, the contraction of the other muscle will lower it.

▼ **Muscles that move the lower-arm bones work in an antagonistic way. One muscle raises the bone and the other muscle lowers it.**

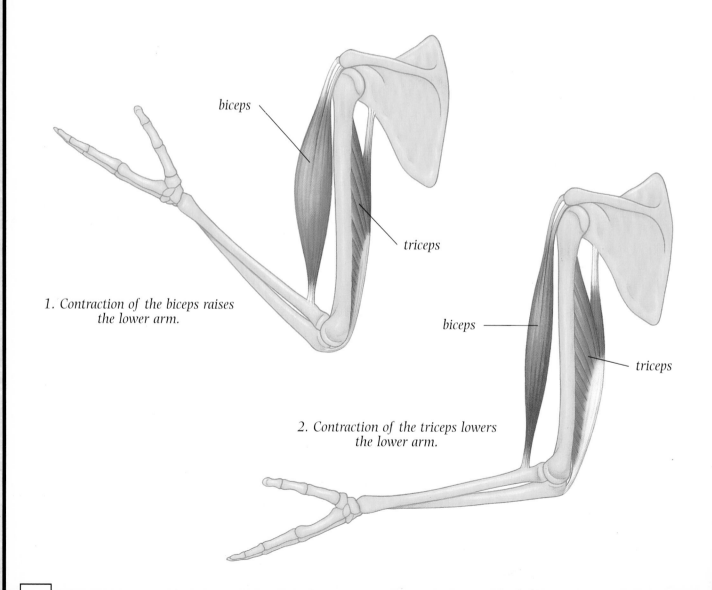

biceps

triceps

1. Contraction of the biceps raises the lower arm.

biceps

triceps

2. Contraction of the triceps lowers the lower arm.

The head balances on top of the spine, pivoting on the top vertebra. Gravity pulls the head forward and down. Muscles at the back of the neck contract to keep the head up. When a person feels sleepy, these muscles relax and the head drops forward —the person "nods off."

The lower arm pivots around the elbow joint. Its movement is controlled by an antagonistic pair of muscles, the biceps at the front of the arm and the triceps at the back. When the biceps contracts, the lower arm is pulled upward and the triceps is straightened. When the triceps contracts, the opposite happens: the lower arm is pulled downward and the biceps is straightened.

The weight of the whole body presses down on the feet. When a person stands on tiptoe, the ball of the foot acts as the pivot. The heel acts as a lever that is lifted when the strong calf muscles at the back of the leg contract. This lifts the weight of the body.

The examples here are simple—one muscle contracting to move one bone in one direction. Many of our activities are more complex, with many muscles involved in moving several bones. Some are large movements, such as the movements of the legs when running. Other activities are small but precise, such as moving the fingers to play the piano. To be able to carry out complex movements like these, muscles have to work together in a coordinated way.

Skeletal muscle contains two types of muscle fiber, white muscle and red muscle. White muscle contracts rapidly, but tires quickly. It is important for short, intense activities such as sprinting. Red muscle contracts more slowly than white muscle, but does not tire so quickly. It is important for activities that require endurance, such as long-distance running.

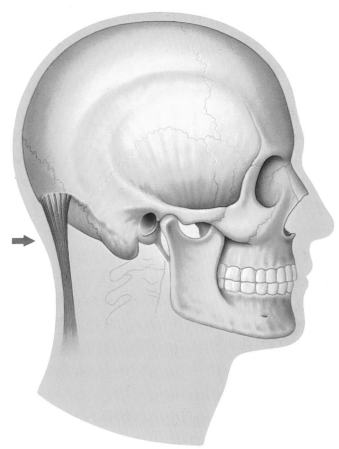

▲ **Muscles at the back of the neck keep the head balanced.**

Muscle Diseases and Problems

Problems with muscles can be the result of activity and exercise or of disease.

Muscular dystrophy is an inherited disease that results in the muscles gradually becoming weaker and eventually wasting away. It can affect the heart muscle as well as the skeletal muscles. It usually affects children and teenagers, but some adults can also be affected. Treatment for muscular dystrophy might include exercise, **physiotherapy**, braces to add strength and, occasionally, surgery.

If a muscle is not used for some time, it gradually becomes weaker and shrinks. This can happen, for example, when a limb has been broken and has had to be encased in plaster for a long time. When the plaster is removed, physiotherapy and exercise are needed to help the muscles in the limb regain their strength. Astronauts who stay in space for a long time also suffer from weakened muscles, because their muscles do not have to work against gravity. They usually follow a strict exercise routine while in space to minimize this effect.

Muscle fatigue is often confused with cramps. It can occur during strenuous exercise, especially in people who are not fit. When a muscle contracts, chemical reactions take place to convert chemical energy into mechanical energy. One of the products of these reactions is lactic acid. Usually, lactic acid is carried to

▲ **Muscular dystrophy is a disease that causes muscles to become weaker and weaker. This child has to use a wheelchair because her muscles are not strong enough to let her walk; she also wears a neck brace to support her head.**

the liver and broken down, but sometimes it is produced more quickly than it can be removed, so it begins to build up in the skeletal muscle. Eventually there is so much lactic acid that the person suffers from muscle fatigue. He or she feels very tired and the muscles ache. When a muscle is in fatigue, it will not contract or relax. It is important not to use a fatigued muscle, because it might begin to use up the proteins that make up its own fibers.

A cramp occurs when a muscle contracts sharply and does not relax. It can be extremely painful. A cramp can usually be eased by gentle massage and flexing of the affected area. There are a variety of causes of cramps, including poor blood flow, a lack of sodium in the diet, or loss of sodium due to heavy sweating.

Tendinitis is an inflammation of a tendon and tendon sheath. It is usually the result of over-exercise of the muscle and can be very painful. The inflammation can usually be reduced by resting the muscles involved.

▼ **Athletes sometimes suffer from muscle fatigue if the muscles are overworked and lactic acid builds up.**

Taking Care of Bones and Muscles

Bones and muscles are very important. Bones provide support, protection, and mobility, while muscles provide the power to move the bones. It makes sense to look after them.

It is important to eat the right kinds of food. Without the right "building blocks," the body cannot grow strong and stay healthy. A sensible diet should include protein-rich foods, such as meat, fish, eggs, nuts, and pulses, which help to build muscles. It should also include foods rich in calcium and vitamin D, such as dairy products and fish, which strengthen bones.

The way we move, sit, bend, lift, and carry weights all have an effect on our muscles and bones. Sitting hunched at a desk is not good for the spine; it is best to keep the spine as straight as possible. Special chairs have been designed that support the spine in the correct position. Slouching in a soft seat such as a sofa, with the spine curved under you, is not a good idea.

◀ **Foods rich in calcium and vitamin D help bones to grow strong and stay healthy.**

Sitting up straight rather than hunching over a desk is better for your spine.

Bend your knees, not your back.

Bending to pick up an object from the ground is often the cause of a back injury. The correct way to lift something is to bend at the knees, keeping the spine as straight as possible. Carrying uneven loads puts stress on the body's framework too. For example, carrying a heavy bag in one hand makes the shoulders lopsided. It is better to distribute the weight evenly: a small bag in each hand or a central knapsack on the back causes much less stress.

▲ **Many back injuries can be avoided by sitting and moving in a way that reduces stress on the spine.**

Professional athletes and other sportspeople are always very careful to "warm up" before they begin their activity. It is important for everyone to do this before exercising, because a warmup helps to avoid muscle strains and other injuries.

▼ **Regular exercise keeps muscles strong.**

Regular exercise helps to keep muscles strong. Muscles that are not used regularly soon become weak and flabby. Heart muscle is the most important muscle in the body, so it is important to do some activity to make the heart work hard. Exercise such as swimming, bicycling, running, and aerobics strengthens the heart muscle and many other muscles too.

Glossary

atlas vertebra The top vertebra in the spine

axis vertebra The second vertebra in the spine

bone marrow Soft tissue at the center of a bone

bursae Sacs containing synovial fluid, which reduce friction inside joints

cartilage Rubbery material that cushions and protects bones

compact bone Strong hard material that makes up one of the outer layers of a bone

contract To get shorter

femur The thigh bone

fracture A broken bone

joints The places where bones meet

lactic acid A waste product made when muscles contract

ligaments Strong fibers that hold joints together

manipulation Gentle movement with the hands

membrane A thin sheet or covering of tissue

minerals Chemicals that are obtained from food in very small amounts. The body needs minerals to stay healthy.

nerves Cells that carry signals from the brain

nutrients Parts of foods that are needed by the body to keep it in good working order

osteo- To do with bones

pectoral girdle The bony structure that forms the shoulders, made up of the clavicles (collarbones) and scapulae (shoulder blades)

pelvic girdle The bony casing around the lower part of the body, made up of the left and right hip bones and the sacrum

physiotherapy The treatment of disease or injury by techniques such as massage or exercise

pulses The edible seeds of certain crops such as peas, lentils, and black beans

radius The bone on the inner side of the lower arm

sesamoid bones Small bones inside joints that help joints to work efficiently

spinal cord The bundle of nerves that runs up the back, protected by the vertebrae

spine The backbone, or spinal column

spongy bone Meshlike material that makes up the center of a bone

synovial joints The most common type of joints. They allow a lot of movement

tendons Bundles of tough fibers that connect muscles to bones

tissue Different types of cells working together to carry out one special function

ulna The bone on the outer side of the lower arm

vertebra One of the bones of the spine

Books to Read

Bailey, Donna. *All About Your Skeleton*. Health Facts. Austin, TX: Raintree Steck-Vaughn, 1990.

Feinberg, Brian. *The Musculoskeletal System*. Encyclopedia of Health. New York: Chelsea House, 1993.

Parker, Steve. *The Skeleton and Movement*. Human Body. Danbury, CT: Franklin Watts, 1991.

Index